Bibliographic information published by the German National Library:

The German National Library lists this publication in the National Bibliography; detailed bibliographic data are available on the Internet at http://dnb.dnb.de .

Imprint:

Print and binding: Books on Demand GmbH, Norderstedt Germany
ISBN: 9783668610132

This book at GRIN:

https://www.grin.com/document/381194

Jennifer Gacich

Connecting Biodiversity Conservation to the Local Community. A Case Study of Organic Agriculture in the Lake Penne Regional Nature Reserve, Abruzzo, Italy

GRIN Verlag

Connecting Biodiversity Conservation to the Local Community

A Case Study of Organic Agriculture in the Lake Penne Regional Nature Reserve, Abruzzo, Italy

Jennifer Gacich

September, 2012

Centre for Development, Environment and Policy (CeDEP)

School of Oriental and African Studies, University of London

Content

Abstract

The aim of this research report is to critically explore stakeholder attitudes towards organic agriculture in the Lake Penne Nature Reserve and to understand whether the reserve's organic agriculture project benefits the local community. Specifically, the perspectives and experiences of cooperative members who manage daily operations in the reserve form an integral part of this research as they are both stakeholders in the protected area and local community members. A case study design is used to conduct an intensive examination of organic agriculture activities in the Lake Penne Nature Reserve, which is a protected area located in the largely rural municipality of Penne, in Abruzzo, Italy. Mixed methods research, including a self-completion questionnaire and unstructured interviews allow the researcher to conduct an in-depth analysis of the potential benefits organic agriculture activities in the protected area bring to the nature reserve and the local community.

1. Introduction

1.1. Aims and objectives

The success of biodiversity conservation is largely dependent on the cooperation of people within and beyond the borders of protected areas. Organic agriculture embodies the concept of biodiversity conservation and community benefits as it is:

> a production system that sustains the health of soils, ecosystems and people. It relies on ecological processes, biodiversity and cycles adapted to local conditions, rather than the use of inputs with adverse effects. Organic agriculture combines tradition, innovation and science to **benefit** the shared environment and promote fair relationships and a good quality of life for all involved.(IFOAM)

The aim of this research report is to critically explore stakeholder attitudes towards organic agriculture in the Lake Penne Nature Reserve and to understand whether the reserve's organic agriculture project benefits the local community. Specifically, the perspectives and experiences of cooperative members, who manage daily operations in the reserve, form an integral part of this research as they are both stakeholders in the reserve and local community members. A case study design is used to 'generate an intensive examination' (Bryman, 2008) of organic agriculture activities in the Lake Penne Nature Reserve, which is a protected area located in the largely rural municipality of Penne, in Abruzzo, Italy. Mixed methods research, using self-completion questionnaires and unstructured interviews allow the researcher to conduct an in-depth analysis of the potential benefits organic agriculture activities in the protected area bring to the nature reserve and the local community.

The *objectives* of the research are to:

1. Understand the background and scope of the project
2. Gain insight into community involvement in the reserve's organic agriculture project
3. Identify the farming community's role in the project
4. Investigate the production and distribution of the organic products

5. Discover how biodiversity conservation knowledge is transferred to the community through the organic agriculture project

Research questions:

- What are the attitudes of stakeholders in the Lake Penne Nature Reserve towards the organic agriculture project?
- How does the organic agriculture project in the reserve benefit the local community?
- Does the community support the production of organic products in the reserve?
- Is biodiversity conservation knowledge transferred to the local community?

The diagram below illustrates how organic agriculture practices in the Lake Penne Nature Reserve can benefit the local community and biodiversity conservation.

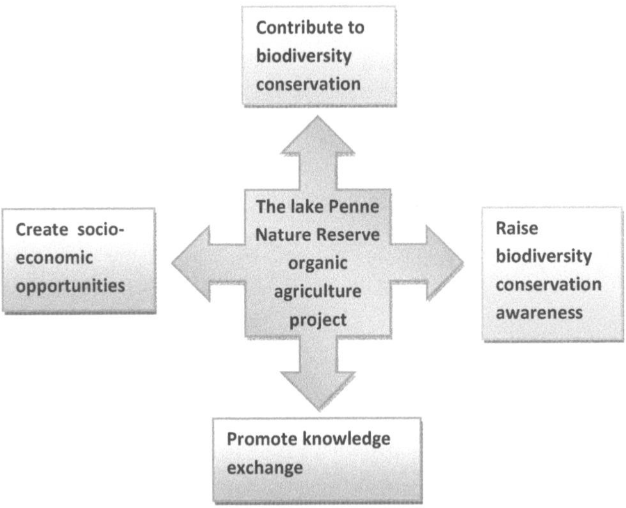

(Author`s own image)

What makes the Lake Penne Nature Reserve organic agriculture project innovative is not only the production of traditional cereals (cultivars) in a protected area, but also the attempt to market reserve made products at the local, regional and national level. Specifically, the cooperative is focussed on making their organically grown spelt, wheat and honey, which are processed and packaged in the reserve, accessible to would be consumers across the country. The findings of this research will reveal some of the difficulties the cooperative has had in gaining access to the domestic organic market as well as issues concerning the demand for organic products.

1.2. Background and context

1.2.1. Organic agriculture in Italy

Agriculture and biological diversity are closely linked. Habitat loss due to land conversion for agricultural activities has drastically reduced biodiversity, affecting 70 percent of all threatened bird species and 49 percent of all plant species (Hilton-Taylor, 2000). However, numerous case studies support the benefits of organic agriculture which can reverse the decline in species diversity, form an integral part of a healthy landscape and improve the socio-economic conditions of farmers (Scialabba et al, 2003).

Organic agriculture has been on the rise in Italy for decades, with the first Italian Organic Agriculture Standards published in 1986 and the founding of a national system for the supervision of regional certification organizations by AIAB in 1988 (IFOAM, 2009). Currently, approximately 8.74 percent of agricultural land is managed organically, with cereals, olives, fruits (grapes, citrus) and vegetables being the main organic products (IFOAM, 2012). Italy was ranked one of the ten European countries with the most organic agricultural land in the 2012 *FiBL-IFOAM Survey on Organic Agriculture Worldwide.* According to the FiBL-IFOAM study, Italy is one of the ten countries worldwide with the most organic producers, with as many as 41, 807 documented producers. However, the distribution of organic sales in Italy is only 3 percent, which is significantly lower than many other European countries such as France, with 8 percent and Germany, with 14 percent (FiBL-IFOAM, 2012). According to the *2007 Bio Bank Report,* the Italian organic market has: 288 purchasing groups, 1,094 specialised shops, 11 organic supermarket brands, 4 online supermarkets, 293 organic restaurants, 15 catering companies, 658 school canteens, 19 organic e-commerce websites, 51 fair trade companies, 135 organic associations, 26 organic exhibitions and fairs and 179 organic books and magazines.

The growth of the organic movement in Italy has primarily been linked to grassroots associations which invested heavily in resources to organize the production, distribution, and promotion of organic products for the domestic market (IFOAM, 2009). Moreover, according to IFOAM, the strength in the organic movement in Italy lies in the ability of the producers to build effective alliances with the growing number of consumers, the environmental movement and international organizations.

1.2.2. The Lake Penne Regional Nature Reserve and Organic Agriculture

The Lake Penne Regional Nature Reserve in Abruzzo, Italy, is a good example of how effective alliances among cooperatives and community members can bolster the production, distribution and promotion of organic products. The nature reserve is overseen by a committee whose members include: Il Comune di Penne (the Penne Municipality), WWF Italia, (WWF Italy), and il Consorzio di Bonifica Centro (an Italian hydropower and water management consortium). Located in a traditionally agrarian landscape, the boundaries of the reserve span over 150 hectares of land, and approximately 1000 hectares of land surrounding the reserve are protected by land use restrictions (APRI, 2001). Visitors to the nature reserve are offered a wide range of services and activities, including:

- A visitor's centre
- An 'environmental education' centre, called "CEA.A. Bellini"
- A natural history museum, called "Nicola De Leone"
- A botanical garden
- An otter centre
- A wildlife area
- Nature trails
- A butterfly garden
- A laboratory and farm
- Accommodation
- A restaurant

Daily operations of the reserve are mainly managed by a locally based cooperative called Cogecstre, whose members are committed to safeguarding the environment. There are also a number of smaller cooperatives operating within the reserve alongside Cogecstre, such as 'Il Gallero', which packages and distributes organic products made in the reserve and 'Samara', which runs the centre for education and training. In the winter months, the education centre holds seminars and workshops, while activities the rest of the year focus on ecotourism, didactic visits and summer camps. The increase in activity during the warmer months allow the cooperative 'Samara', who run the centre in cooperation with Cogecstre, to employ up to nine seasonal workers in addition to their six permanent staff.

The various cooperatives work together to reach the overall management aims of the nature reserve, which are to:

- conserve biodiversity
- preserve landscapes and local heritage
- promote sustainable development and local artisanry
- transfer biodiversity conservation knowledge
- conduct scientific research
- offer services to the local community
- produce, distribute and promote organic agriculture

For the past ten years, the cooperative Cogecstre has invested considerably in organic agriculture research and experimentation. Traditional cultivars such as local 'Abruzzese' spelt and durum wheat have been cultivated on the fertile slopes of 'Collalto Hill', where the soil is largely clay loam and soil pH levels average 7,50 - 8,00 (Cogecstre). Over the years, twelve varieties and ecotypes of spelt have been cultivated on the Collalto Hill farm, including varieties provided by the 'Istituto Sperimentale per la Cerealicoltura', which is an Italian Institute specialized in cereal cultivation and experimentation (Cogecstre). The land set aside for cereal cultivation is managed on a three-year rotation and 'betula', a plant belonging to the birch family, is often used as a cover crop. The images below (*fig. 1 & 2*) illustrate the different land uses on the reserve.

Figure 1. [Di Fabrizio, F. (2011). Cultivation on Collalto Hill [Cogecstre archive online image]. Retrieved August 15, 2012 from http://www.cogecstre.com/

Figure 2. Di Fabrizio, F. (2011). Multiple-use land management on Collalto Hill [Cogecstre archive online image]. Retrieved August 15, 2012 from http://www.cogecstre.com/

Over the years, the demand for organic products made in the reserve has continued to rise steadily and two methods of product distribution have been established by the cooperative. The organic products which are produced and packaged in the reserve are marketed in two forms: (1) the label ***'Sapori Di Campo'*** is used for products sold and distributed locally in the region of Abruzzo; and (2) the Italian World Wildlife Fund label ***'La Terra Del Oasi'*** (meaning 'the land of a protected area'), is used for products sold in WWF affiliated protected areas throughout Italy, on the Italian WWF website www.terredelloasi.it, and through ***Legacoop*** (the largest Italian network of cooperatives). It should be noted that there is a price difference between the two product lines: WWF Terra Del Oasi organic products cost approximately one Euro more than the same products sold through 'Sapori Di Campo'. The price differences between the two product lines are largely due to packaging costs, as Terra Del Oasi products are more elaborately designed and packaged.

The two distribution lines of organic products made through Lake Penne Regional Nature Reserve's organic farming

Figure 3. Di Fabrizio, F. (2011). Spaghetti di farro BIO [Cogecstre archive online image]. Retrieved 15 August, 2012 from http://www.terredelloasi.com/farro-

Figure 4. Di Fabrizio, F. (2011). 'Sapori di Campo' products [Cogecstre archive online image]. Retrieved August 2, 2012 from http://www.cogecstre.com/index1.htm?masseria.htm#masseria

The Lake Penne Regional Nature Reserve is one of five WWF affiliated protected areas involved in the 'Terra Del Oasi' project. The partnership between the cooperative Cogecstre and WWF Italy aims to reinforce alliances among cooperatives, consumers and the community and to engage consumers in conservation efforts which are both sustainable and compatible with the primary aims of biodiversity conservation and ecological management (Stolton & Dudley, 2000). Through the distribution and promotion of organic products made in the reserve, Cogecstre and WWF strive to involve consumers in conservation efforts by demonstrating how the purchase of a single product can contribute to the restoration and maintenance of biodiversity. The shared conviction is that organically cultivated products in protected areas can: conserve biodiversity, benefit local communities, and raise biodiversity conservation awareness. 'Terra Dell'Oasi' organic products carry the WWF logo and are IMC (Mediterranean Certification Institute) certified organic which inform consumers of the products origins, production and greater purpose: the conservation of biodiversity.

2. Literature Review

2.1. Introduction

The topic of organic agriculture and conservation is vast with far reaching political, social, economic and ecological implications. For the purposes of this research report, the scope of this literature review will be limited to: (1). a general overview of how organic agriculture has been defined; (2) organic agriculture in an Italian context; (3) the relationship between communities and organic agriculture. The aim of this literature review is to provide a solid foundation for the development of this research report.

2.2. Defining Organic Agriculture

The shaping and modifying of landscapes in Europe have largely been due to agricultural practices. Modern technology has brought the intensification of agriculture, but at what cost? The intensification of agriculture is associated with the loss of species traditionally associated with agricultural lands. For example, a survey conducted by BirdLife International revealed that farmland habitats account for nearly 60 percent of bird species of European Conservation Concern (Stolton & Geier, 2002). Genetic, species and ecosystem diversity is also eroded with the loss of traditional farming methods and locally adapted varieties. However, sustainable organic agricultural methods have been shown to encourage the expansion of varieties grown, and the preservation of older, locally bred varieties and breeds (Stolton & Geier, 2002). The definition of organic agriculture has been clearly defined by the FAO/WHO Codex Alimentarius guidelines (2007) as:

> 'a holistic production management system which promotes and enhances ecosystem health, including biological cycles and soil biological activity, … [thereby optimizing] the health and productivity of inter-dependent communities of soil life, plants, animals and people.'(p.2)

Farmers who adopt organic agricultural methods employ a series of agronomic practices (e.g. rotations and associations of a large number of plants and animals) which allow organically managed systems to be more biologically complex than those that are conventionally managed (Scialabba et al, 2003). The reliance of organic agriculture on biologically complex agro-ecosystems supports the restoration and maintenance of genetic, species and ecosystem diversity. Therefore, organic agricultural practices are suited to many protected area categories which are dedicated to the protection and maintenance of biodiversity. For example, organizations concerned with the conservation of individual species or habitats such as the Royal Society for the Protection of Birds and the National Trust in the UK are introducing organic systems to the farmed area of their reserves (Stolton & Geier, 2002).

2.3. Organic agriculture and conservation in Italy

One of WWF Italy's main missions is to restore and conserve biodiversity while enhancing landscape quality. Organic farming can contribute to biodiversity conservation by: identifying production systems with ambitious and comprehensive principles for sustainable agriculture; demonstrating that alternative methods of production are commercially and economically viable; and stimulating rapidly growing demand in the marketplace for more sustainably produced products (Vetterli et al, 2002).

Several initiatives linking organic agriculture and biodiversity conservation have been successful in Italy. For example, the Associazione Italiana Agricoltura Biologica (AIAB) project called 'Organic Agriculture and Agroecology in Regional Parks', was designed to promote organic agriculture in the regional parks of Emilia Romagna. The benefits of the project included positively changing local farmers' attitudes towards organic farming, increasing local participation in the project, the provision of training courses and extension services for local farmers, seminars, workshops, farm demonstrations, etc. (Stolton & Geier, 2002). What is not mentioned in the case study is how the organic food products were promoted and distributed. In order to raise conservation awareness (and be economically viable), organic products from protected areas need to reach consumers in the domestic and/or international organic food market.

The market for certified organic products in Italy has seen strong and steady growth over the years. In 2005, its total market value was 2.4 billion €, with the domestic market accounting for 1.7 billion €, exports 0.5 billion €, and public institutions 0.2 billion € (IFOAM, 2009). With over a thousand specialized shops selling organic foods, a growing number of organic retail chains such as 'NaturaSi' and the increasing demand for organic products from both the private and public sector, WWF Oasi and Cogecstre have the opportunity to launch their certified organic products into the domestic market. The question is how to do so effectively. According to IFAOM (2009), the first marketing efforts in Italy were led by farmers, mostly in cooperatives, who faced many legal and political challenges while being underfinanced. Many of these challenges remain and the best methods to overcome them have yet to be identified. One of WWF Oasi's goals is in fact to provide a best practice model for other organizations involved in organic agriculture projects in protected areas so that they may overcome obstacles, establish innovative partnerships and reach consumers willing to partake in conservation efforts.

2.4. Communities and organic agriculture

The growing demand for organic products and the high premiums consumers are willing to pay can also favour local economies. The Mountain Community of Garfagnana has gained from the traditional, synthetic, pesticide free cultivation of spelt wheat, once a Mediterranean staple food, now highly valued by consumers for its high fibre content and healthy properties (Scialabba et al, 2003). To ensure that traditional cultivation methods were being practiced and to increase the value in local production, the Mountain Community of Garfagnana applied for, and obtained, European Recognition for Protected Geographical Indication (PGI) in 1996 (Scialabba et al, 2003). Following its PGI recognition, regulations were drawn up for the Garfagnana spelt common variety *(Triticum dicoccum)* and agronomic practices prescribed for its production as organic (Scialabba et al, 2003). Ultimately, these measures have served to conserve a locally adapted genotype of spelt, guarantee consumers a product which is certified organic and boost the local economy.

Similarly, the management aims of organic agricultural practices in the Lake Penne Regional

Nature Reserve are to conserve local varieties of spelt, provide consumers with certified organic products and benefit the local economy. What is not addressed in the 'Mountain Community of Garfagnana' case is the need to increase consumer awareness of the link between biodiversity conservation and organic agriculture. Some of the ways Cogecstre and WWF Italy are attempting to address this 'gap' are by: (1) generating and publicizing research (Cogecstre also operates a publishing house), (2) distributing and promoting Lake Penne Regional Nature Reserve organic products within the growing Italian organic market, (3) establishing partnerships with local farmers, and (4) providing educational and training opportunities to reserve visitors.

As stated in the *IFOAM Guide to Biodiversity and Landscape Quality in Organic Agriculture* (2009), a wide range of measures exist to enhance biodiversity and landscape quality that may offer multiple benefits, such as the opening of new market opportunities and/or the improvement of farmers' livelihoods and quality of life. Moreover, the IFAOM Guide states that sharing experiences through expanding communication mechanisms can lead to learning and to an ongoing process of further innovation (Bosshard et al, 2009). What is not explored in detail is the mechanisms to disseminate information and experiences. While the importance of organic agriculture and its connection to biodiversity conservation are well-documented, there is still a gap in the transfer of this knowledge to communities and consumers. The partnership among cooperative members in the Lake Penne Regional Nature Reserve and WWF Italy could set an example for other protected area managers who wish to connect people to biodiversity conservation through innovative organic agriculture projects.

3. Methodology

3.1. Mixed methods to answer research questions and generate new theory

A complementary approach to data collection and analysis was needed in order to address the main research questions of this study. Specifically, both qualitative and quantitative research methods were employed to critically explore stakeholder attitudes towards organic agriculture in the Lake Penne Regional Nature Reserve and to understand whether the reserve's organic agriculture project benefits the local community. Methods in data collection and analysis from the two research strategies are employed so that different aspects of the investigation can be dovetailed (Bryman, 2008). The perspectives and experiences of cooperative members form an integral part of this research and mixed methods allow the researcher to 'bring together a more comprehensive account of the area of enquiry' (Bryman, 2008). Although the findings of this case study cannot be generalized to other nature reserves, they can act as a platform on which further study can be conducted. The results of this study can generate a contextual understanding of the opportunities organic agriculture practices in the Lake Penne Regional Nature reserve can create for the local community.

3.2 Research design

The research design of this study is based on the intensive examination of a single organization which is implementing a new strategy to promote organic agriculture and conservation involving the production, packaging and distribution of organic food products made in the Lake Penne Regional Nature Reserve. Mixed research methods offer a more comprehensive account of the area under study, allow a range of techniques to be attempted and reduce the risk of minimal data inputs if one or more methods fail. In addition, qualitative research methods can offer a more in-depth explanation for the findings generated by a quantitative self-completion questionnaire with closed ended questions. The following is an outline of the research design and methods used to collect and analyse data.

3.3. Data collection

To begin with, preliminary discussions with Cogecstre cooperative workers took place in early March, 2010 during on onsite visit to the Lago di Penne reserve to help define relevant questions to be asked in the self-completion questionnaire and during the theme guided interviews. During the visit, some pilot open-ended questions concerning the perceived benefits of the project were asked. This step was essential as translation errors may have occurred from English to Italian and questions needed to be placed in a cultural/historical context in order to avoid missing data. These preliminary questions also helped to generate closed questions and fixed choice answers in the questionnaire. During the visit, key members of the cooperative were asked if they would be willing to participate in the study. They were informed of the following: the purpose of the research; that their participation was voluntary; their right to withdraw from the study at any time; their right not to respond to questions during the process; and the confidentiality of any personal information given.

The self-completion questionnaire was sent to the Nature Reserve in hard copy form via post and electronically, via email. A letter accompanied the questionnaires, outlining the purpose of the study and ensuring full confidentiality. Informal phone interviews were conducted following completion of questionnaires to encourage respondents to reveal further insight into research questions. As respondents had already become familiar with the study, they were more able to comment on topics addressed in the questionnaire. Non-probability convenience sampling was employed to collect quantitative research. Reasons for this sampling method choice include limited access to potential participants and the relatively small number of cooperative workers employed onsite. A self-completion questionnaire was administered containing mostly closed ended questions concerning the perceived benefits of the organic agriculture project in the reserve. Using the 'Likert Scale', questions served to assess participants' level of agreement or disagreement with statements concerning various ecological, social, and economic aspects of organic agriculture practices in the reserve.

For the theme based interviews, participants were selected on the bases of purposive sampling. Key actors relevant to the research questions were asked to participate in phone interviews. The purpose was to gain further insight into perceptions of strengths and weaknesses of the project, community reaction to organic agriculture, future prospects, etc. A snowball sample was attained as social contacts between the researcher and respondents were used to identify additional respondents.

3.4. Data analysis and research constraints

A 'thematic analysis' was used to organize and analyse the qualitative data generated in the five interviews. Detailed notes were taken during the phone interviews which were conducted in Italian (not the researcher's mother-tongue). The excerpts from each interview were used to code the data, which was then grouped into themes and subthemes. The Framework approach allowed the researcher to code, organize and analyze the data so that it could generate a contextual understanding of the opinions expressed in the self-completion questionnaire.

Descriptive statistics are used to make general observations about the data collected (Coakes, Lyndall & Price, 2008). As a Likert Scale is used in the self-completion questionnaire, frequency distributions are displayed in tabular form, bar charts and pie charts. All quantitative data was analysed using the IBM SPSS Version 18 statistical programme. Negatively worded questions concerning the negative impact of the economic crisis in the reserve were also recoded. Statistical tasks that were foreseen in the research design, but not possible due to the small number of respondents included correlation tests, such as a contingency table with the chi-square test to analyse the relationship between the start of the organic agriculture project (independent variable) and the number of visitors to the site or farm demonstrations onsite (dependent variable).

A main limitation of this study however, is the low response rates to the self-completion questionnaire which were not high (13 out of 22 were completed). The low response rate was primarily due to the researcher's inability to access some cooperative members who are not permanently based on the reserve. One fifth of the self-completion questionnaires (in hard copy form) were distributed to cooperative members who 'passed through' the administrative centre of the reserve, but were not returned. Ideally, the researcher would have made many trips to the reserve throughout the year to personally distribute questionnaires to all cooperative members. However, a number of constraints made visits to the reserve difficult, including: limited resources, time constraints, and the distant location of the reserve from Rome, where the researcher is based. Should further research be conducted on the organic agriculture project in the Lake Penne Natural Regional Reserve, such as a longitudinal study on the production, distribution and sales of 'Sapori Di Campo' and Terra Dell'Oasi organic products, the researcher would ideally be locally based in order to access as many cooperative members as possible and to gain a complete picture of the projects' success.

It should be noted that the results contained **no** missing data. This could be an indication of positive respondent interest in the questions as well as the manageable length of the questionnaire (3 pages). Specifically, effort was made by the researcher to avoid repetitive questions and to ensure a variety of topics were covered in a limited number of questions.

4. Results

4.1. General overview of results

The findings of this research are presented using both quantitative and qualitative data results. The aim is to present a comprehensive account of Lake Penne Regional Nature Reserve cooperative member attitudes towards organic agriculture in the reserve. The topics deemed by the researcher to be the most relevant to answering the proposed research questions (see section 3.1) will be presented. The results which the researcher feels would merit further investigation, but go beyond the scope of this research are presented at the end of the section.

An important finding of this research is that general attitudes of cooperative members towards the organic agriculture project in the Lake Penne Regional Nature Reserve are positive. For example, in the self-completion questionnaire, twelve out of thirteen respondents ***strongly agreed*** that organic agriculture safeguards the health of workers; only one respondent selected 'neutral' to this question. Moreover, all respondents to the self-completion questionnaire and telephone interviews ***agreed*** on the following themes:

1. Organic agriculture contributes to biodiversity conservation as well as the preservation of local heritage and landscapes.
2. Organic agriculture practices in the reserve create new socio-economic opportunities for the community.

One of the limitations of this research is the low response rate to the self-completion questionnaire and the difficulties the researcher had in accessing interview candidates. The findings of this research would likely have been somewhat different had more cooperative members working in agriculture completed the questionnaire and/or participated in telephone interviews. The pie chart below (*Figure.5*) illustrates the low number of respondents working in agriculture, education and training, and conservation (one response per sector). However, to ensure validity, it should be noted that many respondents to the self-completion questionnaire and interviewees are long-standing members of the cooperative with extensive knowledge of reserve operations.

Figure 5. **Number of self-completion questionnaire respondents per job sector**

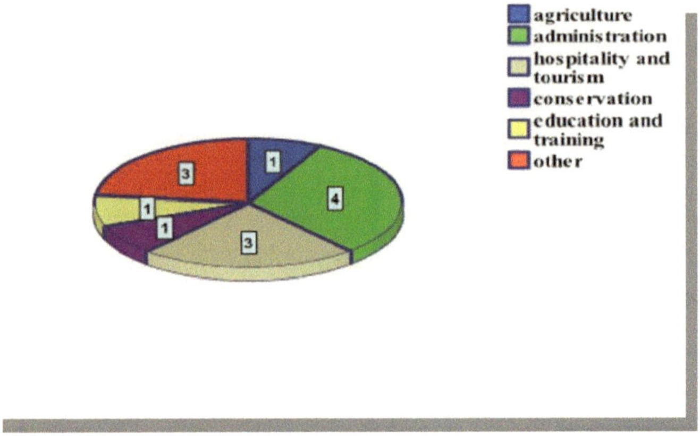

(Author`s own image)

4.2. Consumer willingness to pay more for reserve-made organic products

Another key finding of this research is the divergence in cooperative member attitudes towards consumer willingness to pay higher premiums for organic products made in the reserve. The bar chart below (*Figure. 6*) illustrates the discrepancy numerically.

Figure 6. **Attitudes towards consumer willingness to pay more for organic products made in the reserve**

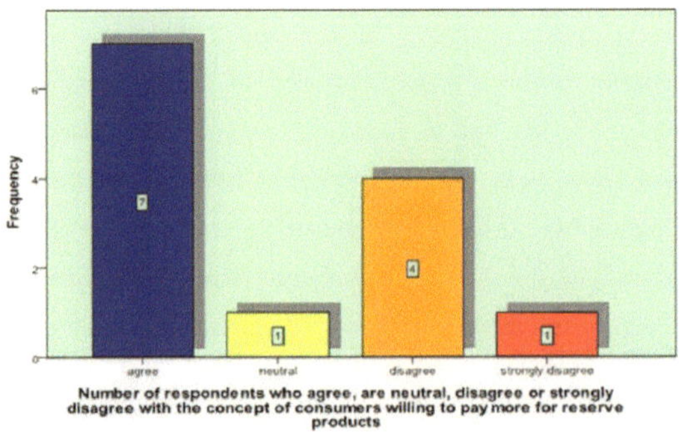

(Author`s own image)

To gain a better understanding of what respondents felt influenced consumer willingness to pay more for organic products, the question was posed systematically to all telephone interviewees. The researcher then identified the recurring sub-themes which influenced participant attitudes towards consumer willingness to pay more for organic products:

- Typical 'working class' families cannot afford to buy organic products on a regular basis
- Low-end supermarkets where many people shop to save money do not offer organic products
- Food consumed in the privacy of homes does not represent a status symbol and is therefore not worth the extra cost
- Many Italian families are no longer used to consuming unrefined grains such as semi-refined spelt
- It is generally believed that children do not like the appearance or taste of semi or unrefined food products, therefore they are not introduced in homes

4.3. Organic agriculture and socio-economic opportunities

A key variable in determining the success of organic agriculture activities in the Lake Penne Regional Nature Reserve is job creation. Currently, the organic agriculture project generates substantial revenue for the reserve. According to the interviewee working in agriculture, Sapori Di Campo products generate approximately 20, 000 Euro in revenue per month. It is the only commercial sector in the reserve which has grown and seen increased revenues over recent years despite the economic recession in Italy. The success of the project to date has allowed for job creation both in the reserve and through partnerships with local farmers who cultivate between 20 to 40 percent of the certified organic spelt used to make Sapori Di Campo products. The positive attitudes of cooperative members towards organic agriculture in the reserve and job creation, are illustrated in the Table (1) below.

Table 1. Attitudes towards organic agriculture and job creation in the reserve

		Frequency	Percent	Valid Percent	Cumulative Percent
Valid	strongly agree	3	23,1	23,1	23,1
	agree	10	76,9	76,9	100,0
	agree	**13**	**100,0**	100,0	

Analysis of the response to the self-completion questionnaire statement, '*The economic recession in Italy has had a negative impact on job creation in the reserve*' produced interesting results. The *majority* of respondents agreed with the statement, which is reflected in the pie chart (*Figure 7*) below.

***Figure 7.* Respondent beliefs that the economic recession in Italy has had a negative impact on job creationin the nature reserve**

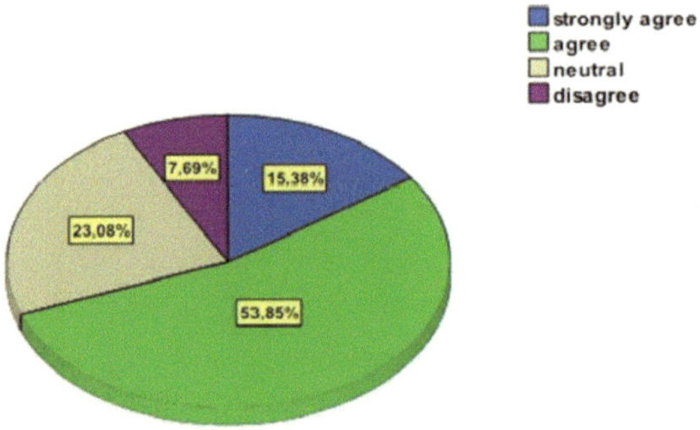

(Author`s own image)

This figure sharply contrasts with the previous Table (1) which shows the same respondents' positive attitudes towards organic agriculture and job creation in the reserve. Discussions concerning job creation in the reserve were conducted in the unstructured interviews to explore the theme further. All respondents agreed that organic agriculture activities in the reserve had *the potential* to create job opportunities. However, a number of constraints were mentioned:

- ❖ Market volatility makes future growth of the project uncertain
- ❖ Competition among domestic organic products exists
- ❖ Stringent organic certification requirements of major organic food distributors hinder distribution expansion
- ❖ Organic certification costs are high for local farmers

4.4. The transfer of biodiversity knowledge

All respondents agree that biodiversity conservation and knowledge is transferred to students who visit the reserve. One of the interviewees, who works in the CEA.A.Bellini education centre, reported that out of the 20, 000 visitors the centre receives annually, 58% were minors (under 18 years of age). Most of the children and young adults visit the reserve with school groups and summer camps. The interviewee highlighted children's enthusiasm for organic products made in the reserve, particularly when they visit the farm and factory where the reserve products are made and interact in the production process (such as making pasta with spelt flour grown in the reserve). The interviewee added that seeing where the organic products originate from, the transformation process and ultimately, the consumption of the products help raise conservation awareness among visitors.

4.5. Farmer access and involvement in the reserve

Every Wednesday, the Lake Penne Regional Nature Reserve invites local farmers to use the reserve's traditional mills. Typically, local farmers bring grains they have cultivated for personal use to the reserve and use the mill to refine their produce (usually into spelt flour). However, opinions as to whether local farmers with access to the reserve can learn and experiment with new techniques in organic agriculture were mixed. Nine out of thirteen respondents agreed that farmers can learn and experiment with innovative techniques in the reserve. However, 2 respondents had a neutral attitude and 2 disagreed with the statement. Conversations in the interviews helped the researcher interpret the findings and place them in a local context. The following themes emerged from the conversations:

- Most farmers who access the reserve are elderly and practise traditional growing methods
- Farmers who gather in the reserve on Wednesdays tend to socialise and share experiences

One respondent whose workshop is located near the mill described the weekly encounters of farmers as 'a Saturday in the village'. A number of respondents reiterated the same concept: the reserve offers local farmers (and thereby the community) and important service as knowledge and experiences can be shared in the context of a nature reserve.

Respondent opinions also varied on the theme of biodiversity conservation knowledge being transferred to local farmers. Only 8 out of 13 respondents agreed with the statement, 'biodiversity conservation knowledge is transferred to local farmers'. However, further analysis of this theme in the unstructured interviews revealed that most respondents associated local farmers with traditional agricultural practices. What farmers seem to be lacking is *'organic certification'* rather than knowledge in organic agriculture cultivation methods. Moreover, most interviewees reported having family members who still cultivate small areas of land using traditional cultivation methods. Most respondents also agreed that local farmers have developed a positive attitude towards organic agriculture practices in the reserve. An interesting factor mentioned by interviewees is that organic agriculture production in the reserve does not create competition among the local farming community. It was said that between 20 and 40 percent of the spelt and durum wheat used by the reserve to make organic products is purchased from local farmers who have organic certification.

The question as to whether the local community is involved in reserve decision making generated interesting results. A total of 6 respondents disagreed with the statement and 4 responded neutrally. However, the decision was made by the researcher not to investigate the issue further as it may have taken the focus away from the main research questions and findings.

5. Discussion

5.1. Overall cooperative member attitudes towards organic agriculture in the reserve

Thorough analysis of both the self-completion questionnaire results and interview outcomes suggest that cooperative members positively view the organic agriculture project in the Lake Penne Regional Nature Reserve. Moreover, organic agriculture in the reserve is seen to benefit the local community by creating socio-economic opportunities.

Organic agriculture production is currently the most profitable commercial activity for the reserve. It is the only sector that has not been affected by the economic crisis and monthly revenue from organic agriculture reaches 20 000 Euros. However, employment opportunities are connected to the future success of the project. An increase in jobs will come with increased distribution and sales of the reserve's products. Currently, the cooperative is undergoing some organizational restructuring as personnel in less productive areas of the reserve will likely be shifted into the more successful ventures. Therefore, one can infer that there is potential for job creation in the reserve's organic agriculture project.

5.2. Contextualizing research findings

The growth in demand for organic products produced in the reserve contrasts with the general decline in conventional agriculture in Italy. According to ISTAT (the main national Italian statistic institute), the overall unemployment rate in the region of Abruzzo was 10,9 percent in the second trimester of 2012. Moreover, there has been a significant decrease in the number workers employed in the agricultural sector in the region. For example, in the territory of Pescara, Abruzzo, the percentage of workers employed in the agricultural sector went from 1, 8 percent in 2006 to 0,8 percent in 2011 (ISTAT). Although the Lake Penne Regional Nature Reserve is located in the municipality of Penne, in the district of Pescara, the production, distribution and demand for organic products made in the reserve have increased over the same time period. This shows that organic agriculture in the reserve is somewhat resistant to the Italian economic downturn, and that it represents a viable commercial activity for the protected area.

As we can see from the results of this study, cooperative members in the Lake Penne Regional Nature Reserve believe that biodiversity conservation awareness can be successfully transferred to the community through various means (education, training, the production and distribution of reserve made organic products, etc.). The cooperative also attempts to involve local farmers in their organic agriculture project, thereby extending the benefits of the project beyond the reserve boundaries.

In the conventional agriculture system, knowledge originates from the top of industry down to the farmer; whereas, in the organic agriculture sector, knowledge is shared and distributed through a chain of actors and organizations, thereby playing a key role in binding networks together (Morgan & Murdoch, 2000). Farmer involvement and access to the Lake Penne Regional Nature Reserve is crucial to the cooperative's ability to create a community network, forge alliances and receive community support. Moreover, in the organic agriculture sector, networks of farmers, cooperatives and distributers can join forces to explore solutions to common problems, such as combining their producer power for marketing purposes (Morgan & Murdoch (2000). The project is viewed positively by the community as the organic agriculture activities in the reserve do not create competition amongst local farmers. Rather, it creates opportunities for some farmers by helping them generate income (through access to the mill or by growing 'Farro' or 'Grano Duro' and selling it back to the reserve).

True to its holistic nature, successful organic agriculture requires those involved to have a broad set of skills, ranging from technical knowledge in areas such as animal husbandry, crop rotation, and pest control to marketing and commercial know-how. Cooperative members in the Lake Penne Regional Nature Reserve work alongside one another to advance the aims of the reserve and all feel they have a stake in the success of the reserve's organic agriculture project, regardless of whether they are directly involved in the project. The alliance between cooperative members and WWF Italy to promote the two lines of organic products made in the reserve, 'Sapori Di Campo' and Terra Dell'Oasi', aim to create opportunities for the transfer of knowledge in organic agriculture and conservation so that it may be spread from cultivators, to communities to the general public to consumers.

6. Conclusion and recommendations

The research questions presented in the introduction of this dissertation were systematically explored and the findings have led the researcher to confirm the hypothesis that organic agriculture activities in the Lake Penne Regional Nature Reserve benefit the local community. Cooperative member attitudes towards the organic agriculture project in the reserve were analysed and a mixed research approach was employed to structure the methodology used for analysis.

The findings of this study reveal the many ways in which organic agriculture activities of the reserve benefit the local community. Participants in the study placed high importance on the potential of the organic agriculture project in the reserve to create socio-economic opportunities for the local community. Efforts are made by the cooperative to forge alliances with local farmers and the local community by encouraging participation in the reserve's organic agriculture activities. Cooperative members also expressed a firm belief that biodiversity conservation knowledge can be successfully transferred to the local community through reserve activities, including organic agriculture.

It is the recommendation of the researcher that further study be conducted on the Lake Penne Regional Nature Reserve organic agriculture project to explore questions raised in this study. A large scale national survey on consumer attitudes towards organic products made in protected areas could be conducted. This would allow the researcher to delve more deeply into themes such as consumer willingness to pay higher premiums for organic products and whether organic products made in nature reserves have additional value. The ability of the cooperative and WWF Italy to gain access to the larger domestic organic food market has yet to be revealed. Lessons learned from the production, distribution and sales of the product lines, 'Sapori Di Campo' and 'Terra Dell'Oasi' could aid other protected area managers develop strategies and implement organic agriculture activities within protected areas. The question remains as to whether organic products made locally in protected areas can grow beyond the niche market they are currently in and expand into the mainstream organic food market.

References

APRI. (2001) *Progetto di rete per le riserve naturali regionali d'Abruzzo. Caso riserva naturale lago di penne "dalla diga alla riserva: come trasformare un vincolo in un'opportunità"*, Cogecstre, Abruzzo, Italy

Bio Bank. (2007) *Bio Bank Report 2007: the new Bio Bank Report on the organic scene in Italy:* [Online] Available from: http://www.biobank.it/en/BIO-articoli.asp?id=729 [Accessed 10th January 2012].

Bengtsson, J., Ahnström., J & Weibull, A. (2005) 'The effects of organic agriculture on biodiversity and abundance: a meta-analysis'. *Journal of Applied Ecology*. Vol. 42 Issue 2, 2005, pp. 261-269.

Bosshard, A., Reinhard, B.R., & Taylor, S. Eds. (2009) *IFOAM Guide to Biodiversity and Landscape Quality in Organic Agriculture.* [Online] Germany, IFOAM. [Online] Available from: http://shop.ifoam.org/bookstore/download_preview/BiodiversityBook-2009-Teaser.pdf [Accessed 5th February 2012].

Bryman, A. (2008) *Social Research Methods, Third Edition.* (New York: Oxford University Press Inc)

Caccamisi, D. (2008) *Sustainable Development of The Regional Park "Sassi Di Roccamalatina" By Agricultural Small-Holders*: [Online] Available from: http://www.ifoam.org/events/ifoam_conferences/owc/modules/abstracts_pdfs/caccamisi_abs_OARWN.pdf

Codex Alimentarius (2007) ISBN 978-92-5-105835-0. *Organically Produced Foods.* 3rd edition. Rome, FAO.

Coakes, S.J., Lyndall, S., & Price, J. (2008) *SPSS: Analysis without anguish: version 15.0 for Windows* (John Wiley & Sons Australia, Ltd)

Cogecstre. (n.d.) '*Masseria Dell'Oasi*' [Online] Available from:

http://www.cogecstre.com/index1.htm?rnr_penne.htm#riserva [accessed 5th July 2012]

Dudley, C., Stolton, S. (2000) The use of certification of sustainable management systems and their possible application to protected area management. In: IUCN/WWF Forest Innovations Project: *Beyond the Trees Conference*, Bangkok, 8-11 May 2000.

FiBL., IFOAM. (2012) *The World of Organic Agriculture* [Online] Available from: http://www.organic-world.net//?id=1675 [Accessed 3rd July]

FiBL., IFOAM. (2011) *The World of Organic Agriculture* [Online] Available from: http://www.organic-world.net/ [Accessed 2nd July]

Geier,B., Stolton, S.(2002) *The relationship between biodiversity and organic farming: defining Appropriate policies and approaches for sustainable agriculture.* [Online] Germany, IFOAM. Available from: http://www.equilibriumresearch.com/upload/document/relationshipbetweenbiodiversityandorganicagriculture.pdf [Accessed 10th September 2012].

Gil J.M., Gracia A & Sánchez M. (2001) *Market segmentation and willingness to pay for organic products in Spain.* [Online] Available from: http://www.unavarra.es/personal/m_sanchez/segmentation.pdf [Accessed 01 September 2012].

Hilton-Taylor, C. (2000) *2000 IUCN Red List of Threatened Species*, IUCN, Gland, Switzerland

IFOAM. (2009) *Growing Organic Information and Resources for Developing Sustainable Organic Sectors; Italy Market Development:* [Online] Available from: http://www.ifoam.org/growing_organic/2_policy/case_studies/italy_market_development.php [Accessed 18th August 2012].

ISTAT. (2012). Occupati e disoccupati. *Principali Indicatori Del Mercato Del Lavoro Per Regione.* II Trimestre 2011 e 2012, pp. 13.

Morgan, K., Murdoch, J. 'Organic vs. conventional agriculture: knowledge, power and innovation in the food chain'. *Geoforum*, 31, 2000, pp. 159-173.

Scialabba, N.E., (2003) Organic agriculture: the challenge of sustaining food production while enhancing biodiversity. In: *United Nations Thematic Group Sub-Group Meeting on Wildlife, Biodiversity and Organic Agriculture*, Ankara, Turkey, 15-16 April 2003.

Scialabba, N. E., Grandi, C., & Henatsch, C. (2003) *Biodiversity and the Ecosystem Approach in Agriculture, Forestry and Fisheries: Case Study No.4, Organic Agriculture and Genetic Resources for Food and Agriculture.* [Online] Available from: http://www.fao.org/DOCREP/005/Y4586E/y4586e05.htm [Accessed 2nd July 2012].

Vetterli, W., Perkins, R. Clay, J. & Guttenstein, E. (2002) Organic farming and nature conservation. In: *OECD Workshop on Organic Agriculture*, Washington D.C., USA 23-26 September 2002.

Wirthen, A. 'Willingness to pay for food produced in accordance with nature conservation criteria: a survey of the food chain'. *Journal on Chain and Network Science*, vol.4, 2004, pp.45-54.

Appendix

Appendix 1. (A) English questionnaire draft and final Italian version sent to cooperative members: Connecting local people to organic farming and biodiversity conservation in the Lago di Penne reserve

INSTRUCTIONS: Please indicate your level of agreement by ticking ONE of the boxes for each question.

1. **Organic farming practices** in the reserve **benefit the local environment** (such as landscape preservation, water management, biodiversity conservation).

☐ 1. Strongly Agree
☐ 2. Agree
☐ 3. Undecided
☐ 4. Disagree
☐ 5. Strongly Disagree

2. **Pesticide and chemical free organic agriculture activities** in the reserve **safeguard** the **health** of **workers**.

☐ 1. Strongly Agree
☐ 2. Agree
☐ 3. Undecided
☐ 4. Disagree
☐ 5. Strongly Disagree

3. The cultivation of **traditional cultivars,** such as spelt and durum wheat contribute to **preserving important elements of local heritage**.

☐ 1. Strongly Agree
☐ 2. Agree
☐ 3. Undecided
☐ 4. Disagree
☐ 5. Strongly Disagree

4. **Knowledge** in **biodiversity conservation** and **organic farming practices** is **transferred** to **local farmers** through outreach programmes, farmer field schools, and access to machinery on site.

- 1. Strongly Agree
- 2. Agree
- 3. Undecided
- 4. Disagree
- 5. Strongly Disagree

5. **Local farmers** with access to the reserve **have developed a positive attitude** towards **organic farming and biodiversity conservation** since the beginning of the WWF COGESCTRE initiative.

- 1. Strongly Agree
- 2. Agree
- 3. Undecided
- 4. Disagree
- 5. Strongly Disagree

6. **Local participation** in **training courses, seminars, and workshops has increased** since the start of the project.

- 1. Strongly Agree
- 2. Agree
- 3. Undecided
- 4. Disagree
- 5. Strongly Disagree

7. Consumers are **willing to pay more** for Lago di Penne **certified organic WWF products** which contribute to local biodiversity conservation.

- 1. Strongly Agree
- 2. Agree
- 3. Undecided
- 4. Disagree
- 5. Strongly Disagree

8. **Knowledge** concerning **organic farming** and **biodiversity** conservation **is transferred to local youth** through scholastic visits to the reserve.

- 1. Strongly Agree
- 2. Agree
- 3. Undecided

4. Disagree

☐ 5. Strongly Disagree

9. **Organic agriculture** activities in the reserve **have lead to job creation** for the local community.

☐ 1. Strongly Agree
☐ 2. Agree
☐ 3. Undecided
☐ 4. Disagree
☐ 5. Strongly Disagree

10. The **economic crisis** in Italy has had a **negative impact** on **the number of visitors** to the reserve.

☐ 1. Strongly Agree
☐ 2. Agree
☐ 3. Undecided
☐ 4. Disagree
☐ 5. Strongly Disagree

11. The recent **economic recession** in Italy has **limited employment opportunities** in the reserve.

☐ 1. Strongly Agree
☐ 2. Agree
☐ 3. Undecided
☐ 4. Disagree
☐ 5. Strongly Disagree

12. **Working conditions** in the reserve are **fair and just** as according to the principles of organic agriculture (IFOAM).

☐ 1. Strongly Agree
☐ 2. Agree
☐ 3. Undecided
☐ 4. Disagree
☐ 5. Strongly Disagree

13. **Local farmers** with access to the reserve **can learn about and experiment with innovative organic farming practices.**

☐ 1. Strongly Agree
☐ 2. Agree
☐ 3. Undecided
☐ 4. Disagree
☐ 5. Strongly Disagree

14. **Organic agriculture** activities within the reserve **can create new opportunities** (market, social, cultural) **for the local community**.

☐ 1. Strongly Agree
☐ 2. Agree
☐ 3. Undecided
☐ 4. Disagree
☐ 5. Strongly Disagree

15. The **local community** is **involved** in the **decision making process** of reserve activities.

☐ 1. Strongly Agree
☐ 2. Agree
☐ 3. Undecided
☐ 4. Disagree
☐ 5. Strongly Disagree

(B) Italian version with an introduction of the research purpose

Lo **scopo** di questo **questionario** è di **capire** se **l'esperienza della Riserva naturale di Penne** ha **prodotto benefici** per la **comunità locale, attraverso** la **salvaguardia del territorio e della biodiversità, lo sviluppo dell'agricoltura biologica** e **delle attività correlate**. I **risultati** saranno **presentati** in **una ricerca** per l'Università di Londra.

Il rapporto della comunità locale con l'agricoltura biologica e la conservazione del territorio e della biodiversità nella Riserva naturale del Lago di Penne

Istruzioni: siete pregati **di indicare se e quanto concordate** con le seguenti affermazioni, tracciando **una sola crocetta** sulla risposta che più rispecchia la vostra opinione.

1. **L'agricoltura biologica produce benefici per l'ambiente naturale locale** (attraverso fattori quali la valorizzazione del paesaggio, la corretta gestione dell'acqua, la conservazione della biodiversità...).

☐ 1. Molto d'accordo
2. D'accordo
☐ 3. Senza opinione
☐ 4. In disaccordo
☐ 5. Completamente in disaccordo

2. **L'attività agricola biologica,** che non prevede l'utilizzo di pesticidi e concimi chimici, **salvaguarda la salute dell'ambiente e degli abitanti** nella riserva e nei territori circostanti.

☐ 1. Molto d'accordo
☐ 2. D'accordo
☐ 3. Senza opinione
☐ 4. In disaccordo
☐ 5. Completamente in disaccordo

3. La coltivazione di **specie tradizionali** come il farro e il grano tenero **contribuisce a preservare** importanti **prodotti del patrimonio locale**.

☐ 1. Molto d'accordo
☐ 2. D'accordo
☐ 3. Senza opinione

☐ 4. In disaccordo
☐ 5. Completamente in disaccordo

4. La **conoscenza** della **conservazione** del territorio e della biodiversità e delle **pratiche dell'agricoltura biologica** è **trasferita** agli agricoltori locali attraverso incontri, seminari, e l'accesso al sito.

☐ 1. Molto d'accordo
☐ 2. D'accordo
☐ 3. Senza opinione
☐ 4. In disaccordo
☐ 5. Completamente in disaccordo

5. Gli **agricoltori locali**, con **l'accesso alla riserva**, hanno **sviluppato** un **atteggiamento positivo** nei confronti dell'agricoltura biologica e della conservazione della biodiversità.

☐ 1. Molto d'accordo
☐ 2. D'accordo
☐ 3. Senza opinione
☐ 4. In disaccordo
☐ 5. Completamente in disaccordo

6. La **partecipazione** agli **incontri** e ai **seminari è aumentata** dall'inizio del progetto.

☐ 1. Molto d'accordo
☐ 2. D'accordo
☐ 3. Senza opinione
☐ 4. In disaccordo
☐ 5. Completamente in disaccordo

7. I **consumatori** sono **disponibili a pagare di più** per **i prodotti bio dell'Oasi** del Lago di Penne **certificati** dall'IMC (Istituto mediterraneo di certificazione) per **contribuire** alla **conservazione della biodiversità locale**.

☐ 1. Molto d'accordo
☐ 2. D'accordo
☐ 3. Senza opinione
☐ 4. In disaccordo
☐ 5. Completamente in disaccordo

8. Con le **visite scolastiche** alla riserva **vengono trasferite ai giovani** la **conoscenza** dell'agricoltura biologica e la **consapevolezza** del **valore** della conservazione del territorio e della biodiversità.

☐ 1. Molto d'accordo
☐ 2. D'accordo
☐ 3. Senza opinione
☐ 4. In disaccordo
☐ 5. Completamente in disaccordo

9. **L'agricoltura biologica** nella riserva ha **creato posti di lavoro** per la **comunità locale**.

☐ 1. Molto d'accordo
☐ 2. D'accordo
☐ 3. Senza opinione
☐ 4. In disaccordo
☐ 5. Completamente in disaccordo

10. La **crisi economica** ha avuto **un impatto negativo** sul **numero di visitatori** nella Riserva naturale.

☐ 1. Molto d'accordo
☐ 2. D'accordo
☐ 3. Senza opinione
☐ 4. In disaccordo
☐ 5. Completamente in disaccordo

11. La recente **recessione economica** ha **ridotto** le **opportunità di lavoro** nella **Riserva naturale**.

☐ 1. Molto d'accordo
☐ 2. D'accordo
☐ 3. Senza opinione
☐ 4. In disaccordo
☐ 5. Completamente in disaccordo

12. Per i **lavoratori** della Riserva naturale **è previsto** un **trattamento equo e giusto**.

☐ 1. Molto d'accordo
☐ 2. D'accordo
☐ 3. Senza opinione
☐ 4. In disaccordo
☐ 5. Completamente in disaccordo

13. Gli **agricoltori locali**, attraverso l'accesso alla riserva, possono **imparare e sperimentare** le **tecniche innovative** dell'agricoltura biologica.

- ☐ 1. Molto d'accordo
- ☐ 2. D'accordo
- ☐ 3. Senza opinione
- ☐ 4. In disaccordo
- ☐ 5. Completamente in disaccordo

14. Le **attività di agricoltura biologica** nella riserva **possono creare** nuove **opportunità** (mercato del lavoro, turismo, attività culturali) per la **comunità locale**.

- ☐ 1. Molto d'accordo
- ☐ 2. D'accordo
- ☐ 3. Senza opinione
- ☐ 4. In disaccordo
- ☐ 5. Completamente in disaccordo

15. Nelle **decisioni** che riguardano le **attività della Riserva naturale** viene **coinvolta** la **comunità locale**.

- ☐ 1. Molto d'accordo
- ☐ 2. D'accordo
- ☐ 3. Senza opinione
- ☐ 4. In disaccordo
- ☐ 5. Completamente in disaccordo

<u>Indicate il settore in cui lavorate:</u>

- ☐ **1.** Agricoltura
- ☐ **2.** Turismo/accoglienza
- ☐ **3.** Ambiente/conservazione
- ☐ **4.** Amministrazione
- ☐ **5.** formazione
- ☐ **6.** Altro

Appendix 2.

Interview Guide Themes for interviews

- **Creation of employment opportunities**
- **Farmers (description and access to reserve)**
- The **business of organic agriculture**/market value/supply and demand/certification
- **Marketing and distribution of 2 products**
- **Obstacles**
- **Taste preference** for refined flour products/price/market/
- **certification**
- **Willingness to pay more for organic products**
- Project **Background**